Juarez Cassiano de Lima Junior
Francisco Dirceu Duarte Arraes
Kleber Gomes de Macêdo

Reference evapotranspiration for the semi-arid region of Ceará

Juarez Cassiano de Lima Junior
Francisco Dirceu Duarte Arraes
Kleber Gomes de Macêdo

Reference evapotranspiration for the semi-arid region of Ceará

Hargreaves and Samani model

ScienciaScripts

Imprint

Cover image: www.ingimage.com

This book is a translation from the original published under ISBN 978-3-330-75514-7.

Publisher:
Sciencia Scripts
is a trademark of
Dodo Books Indian Ocean Ltd. and OmniScriptum S.R.L publishing group

120 High Road, East Finchley, London, N2 9ED, United Kingdom
Str. Armeneasca 28/1, office 1, Chisinau MD-2012, Republic of Moldova, Europe
Printed at: see last page
ISBN: 978-620-8-27084-1

SUMMARY

ACKNOWLEDGMENTS

To all my family, especially my parents, Juarez and Mariana, and my brothers Giancarlo and Caio Victor, for all their education, dedication and affection.

To Anna Hozana for the encouragement, patience and love she gives me every day.

To professors Dr. Francisco Dirceu, Dr. Joaquim Branco and fellow student and friend Kleber Macêdo for their guidance, patience and partnership, which made it possible to publish this book.

To the Federal Institute of Education, Science and Technology of Ceará, Iguatu *campus* for their support.

To the IFCE Geoprocessing Laboratory and all those who make it up.

To my friends Willame Càndido, Gebson Pinheiro, Kleber Macêdo and Wesley Livio, for the jokes and studies we did together on this long academic journey.

What we know is a drop; what we don't know is an ocean.

Isaac Newton

SUMMARY

The Hargreaves and Samani model is used to estimate reference evapotranspiration and is very useful for irrigation management. Its use is practical because it uses simple meteorological elements such as temperature, but it requires regional calibration due to the low yield of its estimates in locations whose characteristics do not resemble those of the place where it was originally established. The aim of this study was to calibrate the Hargreaves and Samani model for semi-arid conditions, using meteorological data from eleven cities in the state of Ceará. To do this, the empirical parameters α (0.0023) and β (0.5) were adjusted, considering two processes: first a simultaneous calibration of the two parameters, and then only the α coefficient. In the simultaneous calibration, the variation reached its extreme values in three cities: α ranged from 0.0004 (Campos Sales and Crateùs) to 0.0019 (Guaramiranga) and the exponent β ranged from 0.51 (Guaramiranga) to 1,1977 (Campos Sales), The correlation between the adjustment and the standard equation showed its maximum value in the city of Jaguaruana, 0.934 for the adjustment with the two parameters and 0.942 for the calibration made with the coefficient α

1 INTRODUCTION

One of the main concerns these days is the amount of water available to meet human consumption, which is increasing as a result of the growth in the world's population and the increase in food production, taking into account that agricultural activities are the most water-demanding for humanity. In this context, the aim of estimating reference evapotranspiration (ETo) is to optimize the use of water in agriculture.

Many models proposed for estimating ETo cannot be applied to scenarios other than those in which they were originally developed. In order to solve this problem, the Food and Agriculture Organization of the United Nations (FAO) proposed the Penman-Monteith (PM) model as the standard for estimating reference evapotranspiration (ALLEN *et al.*, 1998).

PM is a combined equation that expresses ETo considering a hypothetical reference crop with a uniform height of 0.12 m, a fixed surface resistance of 70 s m^{-1} and an albedo of 0.23, growing actively, completely shading the soil and without water restriction. Under these conditions, the only factors that affect ET are climatic elements and, consequently, ETo is a hydrological variable that can be calculated from climatic elements (SENTELHAS *et al.*, 2010).

Penman-Monteith FAO 56 has been defined as the best equation for estimating ETo on a daily and monthly scale in all climates, and can be used globally without any local calibration and even for hourly estimates, due to the incorporation of physiological and aerodynamic parameters, being validated in different environments using precise measurements obtained from Iisimeters.

Its accuracy is high because it has a physical basis, and the calculation process requires reliable measurements of climatic elements such as air temperature, relative humidity, solar radiation and wind speed. However, often not all of these elements are available, due to the limited number of properly equipped weather stations (FERNANDES et al., 2012), or there are major flaws and/or inaccuracies in the measurements. For these reasons, this method's use is limited by the lack of climatic elements.

Allen *et al.* (1998) recommend the use of the Hargreaves-Samani equation (HARGREAVES; SAMANI, 1985) as an alternative for ETo estimation when there is a lack of climatic elements required by the Penman-Monteith FAO 56 model.

When estimating reference evapotranspiration, the HS model only needs simple elements such as minimum, maximum and average air temperatures. This model has been widely used due to its simplicity and the potential for calibrating its parameters (BAUTISTA *et al.*, 2009) and can provide reliable reference evapotranspiration data for the interval of five or more days (TRAJKOVIC, 2007; THEPADIA; MARTINEZ, 2012; JERSZURKI; SOUZA, 2013).

The adaptations to the HS model were not enough to make it independent of the regional calibration of its parameters. In cases where the temperature is low, the HS model overestimates ETo, and in cases where the characteristics of the region do not resemble the one in which its parameters were originally calibrated, the estimate is quite wrong (BAUTISTA *et al.*, 2009; SABZIPARVAR; TABARI, 2010; TABARI; TALAEE,

2011; FERNANDES *et al.*, 2012; SHAHIDIAN *et al.*, 2013).

In a hydrological system, it is also necessary to spatialize the variables that make it up, including ETo (XU *et al.*, 2006). Several authors have used the Geographic Information System (GIS) to spatialize and facilitate the way in which the dynamics between meteorological elements are observed (BARDIN; PEDRO JÛNIOR; MORAES, 2010; CARGNELUTTI FILHO; MALUF; MATZENAUER, 2008; ROCHA *et al.*, 2011). Methodologies for estimating ETo using the Hargreaves and Samani model using digital elevation models were developed by Oliveira *et al.* (2013).

In view of the above, the aim of this study was to calibrate the coefficients of the Hargreaves and Samani (1985) equation and spatialize the ETo estimate for semi-arid conditions in the state of Ceará.

2 LITERATURE REVIEW

2.1 Evapotranspiration

Evaporation and transpiration are phenomena that occur simultaneously and there is no simple way of distinguishing between the two. Evaporation from cultivated soil is mainly determined by the fraction of solar radiation that reaches the surface, this fraction decreasing over the growing season.

In the early stages of the crop, water is lost predominantly through evaporation from the soil, but once the crop reaches full development, covering a greater part of the soil, transpiration becomes the main process. The partitioning of evapotranspiration into evaporation and transpiration is plotted in correspondence to the leaf area per unit of soil surface below it. In sowing, almost 100% of ET comes from evaporation, while in complete vegetation cover, more than 90% of ET comes from transpiration (ALLEN *et al.*, 1998).

2.2 ETo estimation methods

2.2.1 Penman-Monteith

The Penman-Monteith model was adopted by the Food and Agriculture Organization of the United Nations (FAO) as a global standard for estimating reference evapotranspiration, with details presented in FAO 56 (ALLEN *et al.*, 1998). The main limitation to the widespread application of this methodology in irrigation scheduling and management is the time and cost required to acquire and process the necessary meteorological data. There are also concerns about the accuracy of the climate elements observed (DROOGERS AND ALLEN, 2002), as the maintenance of the components of the weather station, specifically instruments such as pyrometers and hygrometers, are often subject to instability and errors (HENGGELER *et al.*, 1996).

2.2.2 Hargreaves and Samani

Analyzing eight years of evapotranspiration obtained from a precision lysimeter and meteorological data from Davis, California, Hargreaves and Allen (2003) observed that 94% of the variance in ETo can be explained by the average temperature, (T_{med}), and the Global Solar Radiation, (R_s).

The Hargreaves and Samani equation is a very simplified empirical model for estimating evapotranspiration, based on climate elements such as Maximum Temperature (T_{max}), Minimum Temperature (T_{min}) and Solar Radiation at the top of the atmosphere (R_a). Although the Hargreaves and Samani model is used to estimate ETo for weekly or longer periods, it can also be used with reasonable accuracy for daily estimates (TRAJKOVIC, 2007; FOOLADMAND *et al.*, 2008) when its coefficients are calibrated locally instead of using the standard values of the coefficients (GAVILAN; CASTILLO-LLANQUE, 2009; ANDRÉ *et al.*, 2010; SABZIPARVAR; TABARI, 2010; RAZIEI; PEREIRA, 2013).

2.3 Parametrization of Equations

Measuring reference evapotranspiration using isometers is often unfeasible due to the high costs and difficulties in handling it. As an alternative, empirical models for estimating ETo have been developed for different locations and methodologies. Although many models have been proposed, there is no single consensus for a given equation in a given climate. Thus, these equations require local calibration (DEHGHANISANIJ *et al.*, 2004). Given the high difficulty of measuring ETo and the high complexity of using the standard PM model, it is necessary to optimize the simplest evapotranspiration estimation models by adjusting them.

The use of a standard method to develop or calibrate an equation to calculate ETo is necessary when measurements of ETo are not available (ALLEN *et al.*, 1994). Irmak *et al.* (2003) cites very significant advantages in calibration, as errors in ETo estimates in poorly calibrated lysimeters can be much greater than errors caused by poor weather conditions and calculations made using data from automatic and/or conventional stations, in addition to the fact that requirements for installing ET measuring lysimeters are very demanding, thus affirming the superiority of calibration based on the PM model when compared to adjustments based on lysimetric measurements. The use of an equation to calibrate or validate another equation is not new and many authors have used and recommended this procedure (ALLEN *et al.*, 1998; ITENFISU *et al.*, 2003; IRMAK *et al.*, 2003).

Fernandes *et al.* (2012) calibrated the coefficients of the Hargreaves and Samani model for the state of Goiás, using climatic elements from ten automatic weather stations.

Borges and Mediondo (2007), for the conditions of São Paulo, observed that the calibration of the empirical Hargreaves coefficient for the rainy season and the dry season provided more accurate estimates of reference evapotranspiration.

Working in desert conditions and with fourteen years of data from synoptic stations in the southern provinces of Iran, Fooladmand and Haghighat (2007) also adjusted the value of the empirical coefficient of the HS model.

In a semi-arid region of Iran (SEPASKHAH; RAZZAGHI, 2009), California and Bolivia (SHAHIDIAN *et al.*, 2013) found very satisfactory correlation values between the Penman-Monteith and Hargreaves and Samani models with the locally calibrated radiation parameter (Kt).

In northeastern Spain, in similar studies, the calibrated α value also fluctuated, in calibrations made with data taken from eight automatic weather stations and one conventional one (MARTINEZ-COB; TEJERO-JUSTE, 2004).

2.4 Solver Supplement

The parameterization of the coefficients of the HS equation became feasible with the Solver application, which is a tool in the Microsoft Excel software®, which is part of the Microsoft Office package®. With Solver, you can find an ideal value (maximum, minimum or any value you choose) for an equation in a cell - called the target cell - according to constraints, or limits, on the values in other cells of an equation in a spreadsheet.

According to Luo and Xie (2010) the solver works with a group of cells, called decision variables or simply variable cells, which are directly and indirectly linked to the calculation of the equations in the objective cells, adjusting the values in the decision variable cells according to the pre-established constraints.

Solver has been used successfully in hydraulic and hydrological modeling, Fontane (2001) used it to solve optimization problems for water supply reservoirs and power generation. Huddleston *et al.* (2004) applied Solver to the analysis of water distribution networks, and the results showed that the application of this technology is an efficient way to solve problems in a relatively complex engineering system, minimizing the computational load.

3 MATERIAL AND METHODS

The study was carried out using meteorological data from the state of Cearà, which is located in the northeast of Brazil, between parallels 2.5° and 8° South latitude and meridians 37° and 42° West longitude, with a total area of 146,348.6 km^2 . According to Koppen's climate classification, the region has three types of climate: BSw'h', Aw' and Cw', with approximately 80% of the area being BSw'h' (hot semi-arid).

Data from conventional weather stations in twenty municipalities were used (Table 1 and Figure 1), and data that are not from the state of Cearà were only used for interpolation when making the spatialization maps. Data from the National Institute of Meteorology (INMET) are: wind speed (u), sunshine (n), maximum temperature (T_{max}), minimum temperature (T_{min}), average temperature (T_{med}), and relative humidity (RH).

Table 1 - Location of the weather stations used in the study

Stations	UF	Latitude (°)	Longitude (°)	Altitude (m)
Acaraù	EC	-2,88	-40,14	16,50
Barbalha	EC	-7,31	-39,30	409,00
Campos Sales	EC	-7,00	-40,38	583,50
Crateùs	EC	-5,16	-40,66	296,82
Guaramiranga	EC	-4,28	-39,00	870,67
Iguatu	EC	-6,36	-39,29	217,60
Jaguaruana	EC	-4,87	-37,76	11,71
Morada Nova	EC	-5,11	-38,36	43,62
Quixeramobim	EC	-5,16	-39,38	79,50
Sobral	EC	-3,73	-40,33	109,62
Tauà	EC	-6,00	-40,41	398,77
Apodi	RN	-5,60	-37,80	150,00
Cabrobó	PE	-8,51	-39,33	341,46
Ouricuri	PE	-7,90	-40,04	459,28
Esperantina	PI	-3,90	-42,25	87,05
Parnaiba	PI	-3,80	-41,76	79,50
Paulistana	PI	-8,13	-41,13	374,22
Picos	PI	-7,03	-41,48	207,93
Piripiri	PI	-4,26	41,78	161,12
Sao Gonçalo	PB	-6,75	-38,21	233,06

Table 2 - Characteristics of the weather stations used, T_{med} - Maximum, average and minimum air temperature, in degrees Celsius; U_{10} - Wind speed (anemometer at a height of ten meters).

Stations	UF	T_{max}(°C)	T_{med}(°C)	T_{min}(°C)	U10 (m s)$^{-1}$	Period
Acaraù	EC	32,05	27,60	23,14	4,20	1976/2011
Barbalha	EC	32,08	26,43	20,78	1,85	1973/2014
Campos Sales	EC	30,93	25,37	19,80	4,44	1962/2014
Crateùs	EC	33,16	27,55	21,95	2,91	1961/2014
Guaramiranga	EC	25,56	21,77	17,97	2,63	1961/2014
Iguatu	EC	33,11	27,68	22,25	2,48	1961/2014
Jaguaruana	EC	33,38	28,14	22,90	3,73	1969/2014
Morada Nova	EC	34,07	28,24	22,41	3,00	1962/2014
Quixeramobim	EC	33,02	27,91	22,80	3,27	1961/2014
Sobral	EC	33,84	28,03	22,22	2,07	1961/2014
Tauà	EC	32,49	26,93	21,38	2,55	1964/2014
Apodi	RN	33,91	28,55	23,18	3,21	1962/2014
Cabrobó	PE	32,12	26,95	21,78	3,70	1963/2014
Ouricuri	PE	31,77	26,46	21,15	2,85	1975/2014

Esperantina	PI	34,03	27,94	21,85	1,74	1979/2014
Parnaiba	PI	31,76	27,37	22,97	4,44	1971/2014
Paulistana	PI	32,89	27,54	22,20	3,52	1975/2014
Picos	PI	34,21	28,20	22,20	2,10	1974/2014
Piripiri	PI	34,25	28,19	22,14	2,14	1976/2014
Sao Gonçalo	PB	33,13	27,00	20,88	2,17	1961/2014

Figure 1 - Location of weather stations.

The FAO recommends that empirical methods should be calibrated and validated using the Penman-Monteith equation as a reference (ALLEN *et al.*, 1998). This equation has shown excellent performance in estimating ETo in a variety of climatic conditions, which is why it was decided to use the Penman-Monteith FAO 56 (PM FAO-56) method as a reference for evaluating, calibrating and validating the Hargreaves equation in this work. The PM FAO-56 equation can be written as (ALLEN *et al*, 1998).

$$ET_o = \frac{0{,}408\Delta(R_n - G) + \gamma \frac{900}{T_{med} + 273} u_2(e_s - e_a)}{\Delta + \gamma(1 + 0{,}34 \cdot v_2)} \quad (1)$$

In which:

ETo - reference evapotranspiration, mm day ;$^{-1}$

Rn - total net radiation of the lawn, MJ m^{-} 2 day ;$^{-1}$

G - soil heat flux density, MJ m^{-2} day ;$^{-1}$

Tmed - average daily air temperature, °C;

u2 - daily average wind speed at 2 m height, m s ;$^{-1}$

es - vapor saturation pressure, kPa;

ea - partial vapor pressure, kPa;

es-ea - vapor saturation deficit, kPa;

Δ - slope of the vapor pressure curve at the Tm point, kPa°C ;$^{-1}$

γ - psychrometric coefficient, kPa°C .$^{-1}$

The slope of the vapor pressure curve is calculated as a function of the temperature.

$$\Delta = \frac{2504 \exp\left(\frac{17{,}27 . T_{med}}{T_{med} + 237{,}2}\right)}{(T_{med} + 237{,}3)} \quad (2)$$

The average monthly temperature (Tmed) is obtained as a function of the maximum temperature (Tmax) and the minimum air temperature (Tmin).

$$T_{med} = \frac{T_{max} + T_{min}}{2} \quad (3)$$

Atmospheric pressure (P) is given as a function of local altitude (z).

$$P = 101{,}3 . \left(\frac{293 - 0{,}006z}{293}^{5,26}\right) \quad (4)$$

The vapor saturation pressure (es) is calculated as a function of the air temperature saturation pressure (e°):

$$e_s = \frac{e^0(T_{max}) + e^0(T_{min})}{2} \quad (5)$$

In which:

$$e^0(T_{med}) = 0{,}6108\exp\left(\frac{17{,}27.T_{med}}{T_{med}+237{,}3}\right) \quad (6)$$

Substituting T_{med} in the equation for T_{min}, we obtain the saturation pressure of the minimum air temperature and° (T_{min}) and T_{med} by T_{max} we get eo (T_{max}).

Current vapor pressure is calculated as a function of relative humidity (RH) and steam saturation pressure (e_s).

$$e_a = \frac{UR}{100}.(e_s) \quad (7)$$

The radiation balance (R_n) is the difference between the shortwave radiation balance (R_{ns}) and the longwave radiation balance (R_{nl}). Obtained from the equations:

$$R_n = R_{ns} - R_{nl} \quad (8)$$

$$R_{ns} = (1-\alpha).R_s \quad (9)$$

$$R_{nl} = \sigma\left[\frac{T_{max}.K^4.+T_{min}.K^4}{2}\right]\left(0{,}34-0{,}14\sqrt{e_a}\right)\left(1{,}35\frac{R_s}{R_{so}}-0{,}35\right) \quad (10)$$

$$R_s = R_a\left(0{,}25+0{,}5\frac{n}{N}\right) \quad (11)$$

$$R_{so} = \left(0{,}75+2\times10^{-5}z\right)R_a \quad (12)$$

$$R_a = \frac{24(60)}{\pi}G_{sc}d_r\left[\omega_s sen(\varphi)sen(\delta s + cos(\omega os(\omega\delta)sen(\omega_s)\right] \quad (13)$$

$$d_r = 1+0{,}033.\cos\left(\frac{2\pi}{365}j\right) \quad (14)$$

$$\delta = 0{,}4093 sen\left(\frac{2\pi}{365}j-1{,}39\right) \quad (15)$$

$$\omega = \cos^{-1}(-\tan\varphi . \tan\delta) \tag{16}$$

$$N = \frac{24.\omega_s}{\pi} \tag{17}$$

In which:

R_a - solar radiation at the top of the atmosphere, MJ m^{-2} d ;$^{-1}$

R_s - solar radiation, MJ m^{-2} d ;$^{-1}$

R_{so} - global solar radiation for clear days, MJ m^{-2} d ;$^{-1}$

N - number of hours of sunshine, h;

N - astronomical length of day, h;

σ - Stefan-Boltzmann constant, $4.903.10^{-9}$ MJ K^{-4} m^{-2} d ;$^{-1}$

$TmaxK^4$ - maximum air temperature, K=°C+237.16;

$TminK^4$ - minimum air temperature, K=°C+237.16;

J - Julian day, day;

δ - solar declination, radians;

Gsc - solar constant, 0.0820 MJ m^{-2} min ;$^{-1}$

dr - relative distance earth-sun, dimensionless;

ωs - hour angle of sunrise or sunset, radians;

φ - local latitude, radians.

The ground heat flux (G) is calculated as a function of the temperature of the month

(T_i) and the temperature of the previous month (T_{i-1}):

$$G = 0,14(T_1 - T_{-1}) \tag{18}$$

Hargreaves and Samani (1985) proposed the following equation for estimating ETo using only the values of maximum, minimum and average air temperature and extraterrestrial radiation:

$$ETo = \alpha(T_{max} - T_{min})^{\beta}(T_{med} + 17,8)Ra \times 0,408 \tag{20}$$

In which:

a - empirical parameter, using its original value of 0.0023; β - exponential empirical parameter, using its original value of 0.5.

Monthly ETo averages were calculated for both methodologies, PM and HS, for all the sites studied. For calibration, the monthly ETo averages for the odd months were used, calculating the ETo averages with the data from the even months, thus validating the calibration of the empirical parameters. Two calibrations were carried out, the first adjusting the α and β parameters simultaneously, and a second calibrating only the α parameter, with β remaining at its original value of 0.5 in order to observe the importance of one parameter over the other. Both calibrations used the same amount of data, which in turn was processed using the same procedures.

The adjusted coefficients and ETo estimates were spatialized with both calibrations, and the estimates were compared with the original Hargreaves and Samani equation (1985). The same procedures employed by Oliveira *et al.* (2013) were used, using a multiple linear regression model with altitude, longitude and latitude as independent variables and measured temperature as the dependent variable based on the following quadratic model:

$$T_{med} = A_0 + A_1.h + A_2.h^2 + A_3.\lambda + A_4.\lambda^2 + A_5.\varphi + A_6.\varphi^2 + A_7.h.\varphi + A_8.h.\lambda + A_9.h.\lambda \qquad (21)$$

In which:

Tmed - Average monthly temperature (i = 1, 2, 3..., 12);

λ - longitude in degrees;

φ - latitude in degrees;

h - altitude in meters;

An - regression coefficients.

The coefficients obtained were applied to the spatialization of ETo in the form of maps on a monthly scale using Idrisi Andes© software, using the same methodologies employed by Oliveira *et al.* (2013).

The parameters of the Hargreaves and Samani (1985) equation were adjusted using Microsoft Excel® using the methodology described by Wraith and Or (1998) for adjusting non-linear equations using the Solver application.

Data analysis was carried out according to the statistical indices suggested by Legates and Mccabe Jr (1999): Willmontt index (id), Nash-Sutcliffe coefficient (E), root mean square error (RMSE), maximum absolute error (EMAX), and ratio between means in percentage (ξ) equations 22 to 26 respectively.

$$id = 1 - \left[\frac{\sum_{i=1}^{N} (P_i - O_i)^2}{\sum_{i=1}^{N} (|P_i - O| + |O_i - O|)^2} \right] \quad (22)$$

$$E = 1 - \frac{\sum_{i=1}^{N} (Y_i - X_i)^2}{\sum_{i=1}^{N} (X_i - X)^2} \quad (23)$$

$$RMSE = \sqrt{\frac{\sum_{i=1}^{N} (Y_i - X_i)^2}{n}} \quad (24)$$

$$EMAX = \max \{ |Y_i - X_i| \}_{i=1}^{N} \quad (25)$$

$$\xi = \frac{\overline{Y_i}}{\overline{X_i}} \quad (26)$$

In which:

H_i - values estimated by the standard model;

P_i - values estimated by the proposed models;

X - average of the values estimated in both models.

Yi - average values obtained by the proposed models;

Xi - average values obtained by the standard model.

4 RESULTS AND DISCUSSION

4.1 PARAMETERIZATION

In the simultaneous calibration of the parameters of the Hargreaves and Samani equation, the variation reached its extreme values in three cities, the α coefficient oscillated from 0.0004 in Campos Sales and Crateùs, to 0.0019 in Guaramiranga, and the β exponent oscillated from 0.51, again in the city of Guaramiranga to 1.1977 in Campos Sales (Table 3 and Figure 2).

According to Fooladmand and Haghighat (2007), locations with wind speed values greater than two meters per second require the use of adjusted coefficients instead of the original, which explains the high disparity of the corrected values in the simultaneous calibration for the cities of Campos Sales, Crateùs and Acaraù, with speeds of 4.44; 2.91 and 4.20 m s^{-1} respectively (Table 2). According to Hargreaves and Samani (1985), the model studied and calibrated does not adapt well to coastal regions, which possibly explains this disparity in the city of Acaraù, a behavior similar to that observed by Valero-Maestro *et al.* (2013) observed calibrating the Hargreaves and Samani equation in a semi-arid region of Spain, using data from 12 automatic stations. On the coast, the adjusted coefficient reached 0.0034, while in more inland regions it did not exceed 0.0025.

In Mexico, a similar behavior was observed on the coast, but this time the greatest difference did not exceed 0.0026 for the months of April, May, October and December, since monthly calibrations were made (BAUTISTA *et al.*, 2009).

The alphas and betas calibrated simultaneously showed inversely proportional behavior, the higher the alpha the lower the beta, and the lower the beta the higher the alpha. The adjustment made only with the alpha coefficient behaved similarly to the alpha of the simultaneous adjustment, only with less disparity between the calibrated alphas, highlighting the importance of adjusting the β parameter, valuing the calibration of the exponent that has an influence on the thermal amplitude.

Table 3 - Empirical coefficient values for the simultaneous and single parameter calibration processes.

Cities	α	β	α'
Acaraù	0,0007	1,03	0,0023
Barbalha	0,0013	0,68	0,0021
Campos Sales	0,0004	1,19	0,0025
Crateùs	0,0004	1,15	0,0024
Guaramiranga	0,0019	0,51	0,0019
Iguatu	0,0008	0,95	0,0023
Jaguaruana	0,0009	0,91	0,0024
Morada Nova	0,0012	0,73	0,0023
Quixeramobim	0,0008	0,97	0,0026
Sobral	0,0011	0,74	0,0020
Tauà	0,0008	0,92	0,0022

Also in the calibration made with the two parameters, the value that differed least from the original value for α was found in Guaramiranga (0.0019), lower even than the 0.0022 found by Bautista *et al.* (2009) in Mexico, for the months of June and July. As for the β coefficient, in the city of Guaramiranga the change was practically nil (0.51) (Table 3). In the municipality of Barbalha, the adjustments reached values of 0.0013 for α and 0.68 for β, values very close to those found in local and regional calibrations made by Fernandes *et al.* (2012), for the state of Goiás, using data from ten weather stations with different time intervals between stations, obtaining 0.00141 as the α coefficient and 0.68 as the empirical β exponent, regionally calibrated.

Calibration with just one parameter, α' (Table 3), ranged from 0.0019 to 0.0026 for the cities of Guaramiranga and Quixeramobim, respectively. The lower value found was to be expected given the first adjustment made, since the exponent β hardly differed from the original. Values found in studies carried out using data from provinces in southern Iran are similar to the results obtained in this study. The annual calibrations carried out by Fooladmand and Haghighat (2007) ranged from 0.0018 to 0.0032 for the original coefficient of the Hargreaves and Samani equation.

In adjusting α', the calibration that differed least from the original value was calculated for the cities of Acaraù, Iguatu and Morada Nova, remaining with its original constant (0.0023). In Jacupiranga - SP, Borges and Mediondo (2007) used eight different methods to estimate ETo, with the correlation made between the Hargreaves & Samani and PM models. Samani and PM models, the empirical coefficient found for HS was 0.00224 for the months of September to April, and 0.00202 for the months of May to August.

When evaluating the adjustments to the HS equation, the calibration made with the two parameters (Table 4) obtained slightly better results than the adjustment made with α' alone (Table 5). The agreement of the proposed model in relation to the PM equation was greater than 0.9 in almost all the cities for the calibration with the two parameters, except for the stations located in the municipalities of Acaraù, Campos Sales and Tauà, where the Willmott index was 0.833; 0.882 and 0.837, respectively (Table 4). These values are similar to those found by Bautista *et al.* (2009) in Mexico. In the calibration with only the α coefficient, the municipalities with the highest Willmontt indices were Guaramiranga (0.936), Jaguaruana (0.905) and Morada Nova (0.926). The other values are higher than those found for the city of Uberaba - MG, whose Id was only 0.7 (MELO *et al.*, 2012).

Table 4 - Evaluation of calibration performance with α and β.

Cities	Id	E	RMSE	EMAX	r^2
Acaraù	0,830	0,580	0,641	1,772	0,770
Barbalha	0,900	0,650	0,485	1,408	0,820
Campos Sales	0,880	0,640	0,805	2,270	0,800
Crateùs	0,940	0,790	0,582	2,776	0,890
Guaramiranga	0,940	0,780	0,289	0,849	0,890
Iguatu	0,950	0,820	0,466	1,650	0,910
Jaguaruana	0,960	0,860	0,405	1,048	0,930
Morada Nova	0,960	0,850	0,388	1,154	0,920
Quixeramobim	0,950	0,810	0,579	1,835	0,900
Sobral	0,930	0,760	0,52	1,554	0,880
Tauà	0,840	0,490	0,714	1,893	0,720
Average	0,920	0,730	0,534	1,655	0,860

In the analysis of the square root of the mean error (RMSE) for the calibration made with the two parameters, the values found ranged from 0.289 (Guaramiranga) to 0.805 (Campos Sales), while Acaraù showed a value higher than that found by Bautista *et al.* (2009), in adjustments made with data from coastal and continental regions in Mexico, RMSE equal to 0.43 and 0.42 respectively.

Table 4 - Evaluation of calibration performance using the α' parameter

Cities	Id	E	RMSE	EMAX	r^2
Acaraù	0,709	0,469	0,721	1,643	0,788
Barbalha	0,879	0,623	0,507	1,451	0,791
Campos Sales	0,757	0,477	0,966	2,298	0,713
Crateùs	0,847	0,653	0,756	2,067	0,874
Guaramiranga	0,936	0,780	0,289	0,854	0,884
Iguatu	0,896	0,722	0,581	1,936	0,870
Jaguaruana	0,905	0,762	0,528	1,369	0,942
Morada Nova	0,926	0,785	0,467	1,309	0,901
Quixeramobim	0,800	0,442	0,959	3,827	0,729
Sobral	0,894	0,725	0,559	1,571	0,885
Tauà	0,799	0,524	0,689	1,617	0,733
Average	0,850	0,633	0,638	1,813	0,828

The adjustment made to the α' coefficient (Table 5) showed the importance of the β exponent in the HS model. Compared to the adjustment made with the two parameters (Table 4), the indices of agreement and efficiency were lower, and the calculated errors were higher; the correlation between the adjusted model and the standard equation (PM) was good ($r > 0.80$) in almost all locations, except for Acaraù, Barbalha, Jaguaruana, Sobral and Tauà. Also according to Tables 4 and 5, Guaramiranga showed a unique behavior: in the adjustment made with the two parameters, the β exponent practically did not differ from its original value; as a result, the two calibrations behaved in the same way and showed the same statistical results.

Comparisons between the models adjusted with the standard (PM) and the original Hargreaves and Samani equation (Figures 2 and 3) clearly show the variability in ET estimates from one region to another. In practically all the stations, the original HS equation overestimated the ETo calculated by PM in the first few months of the year, the period with the highest rainfall and lowest temperature variations.

The Hargreaves and Samani (1985) model tends to overestimate ETo in humid regions and underestimate it in very dry regions and in regions with higher wind speeds (TEMESGEN; ALLEN; JENSEN, 1999; XU; SINGH, 2002). Therefore, the Hargreaves and Samani (1985) model requires local calibration before it can be applied to estimate ETo in any specific region (BAUTISTA *et al.*, 2009; GAVILAN *et al.*, 2006; FOOLADMAND; HAGHIGHAT, 2007).

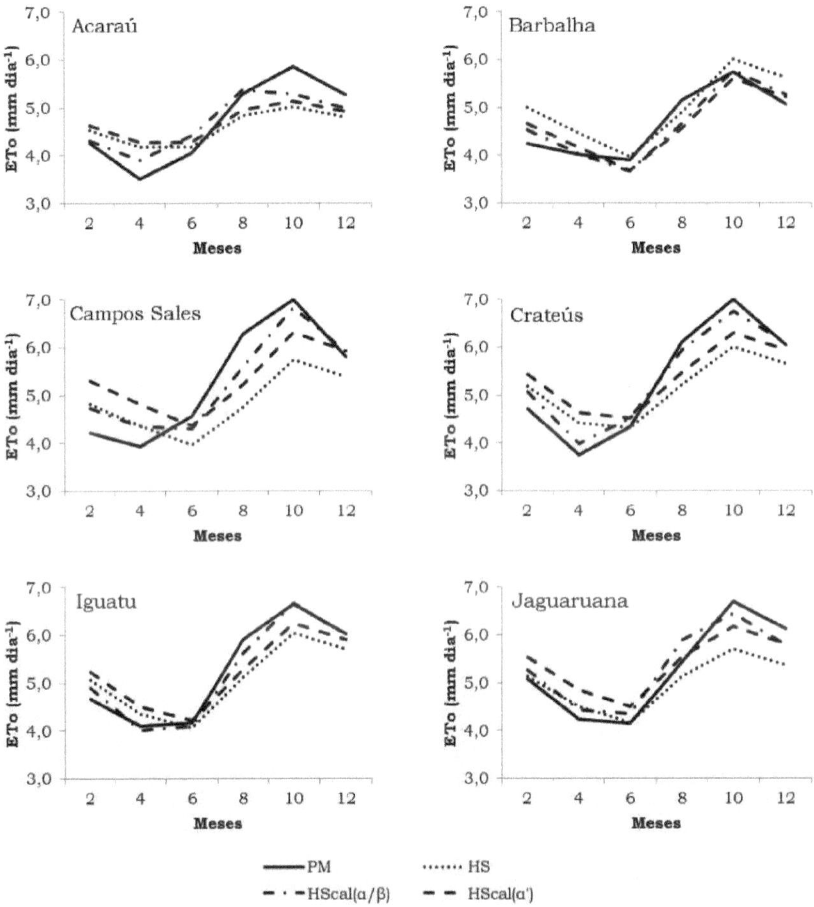

Figura 2 - Comparison of calculated ETo values by PM, HS and adjusted models for Acaraù, Barbalha, Campos Sales, Crateùs, Iguatu, Jaguaruana.

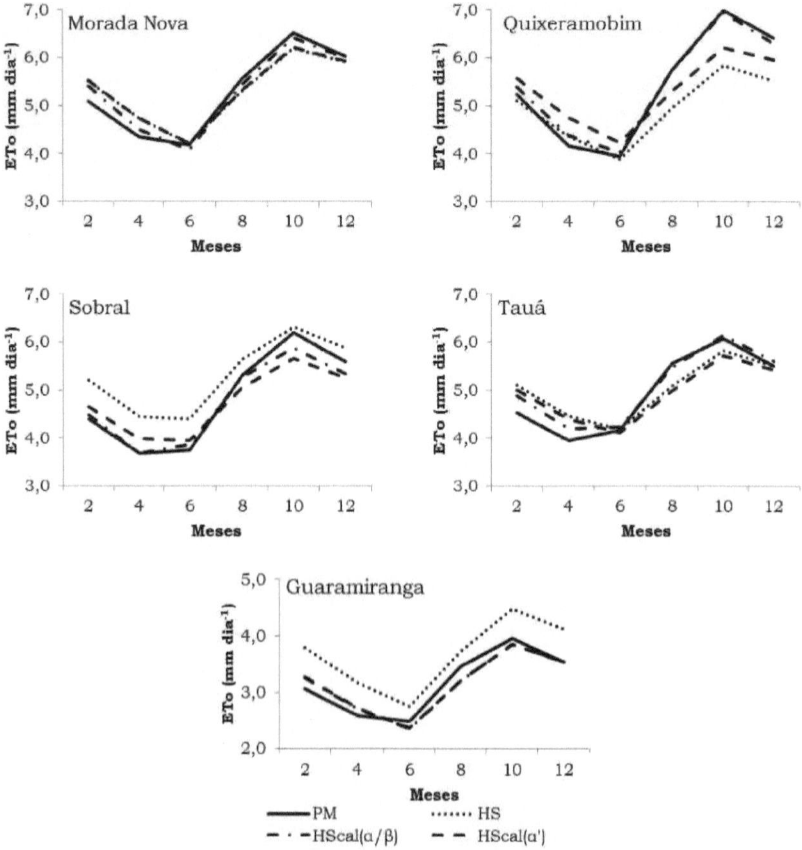

Figura 3 - Comparison of calculated ETo values by PM, HS and adjusted models for Morada Nova, Quixeramobim, Sobral, Tauà and Guaramiranga.

It can also be seen that the model with only the α' parameter adjusted overestimates PM in the first few months of the year and tends to underestimate it from the beginning of the second half of the year, showing greater differences than the model with two parameters adjusted, and sometimes even with a greater error than the original model (Figure 2 and 3). Most of the stations showed satisfactory results throughout the year, with values close between the adjusted models and the standard equation. Talaee (2014), comparing six versions of the HS equation adjustments, noted that the model that performed best in the semi-arid climate was the one calibrated in Davis, California, a city with a hot, dry climate, which fits the original model well.

The values that differed most when comparing the original HS equation with the standard PM model were in the cities of Acaraù, Campos Sales and Crateùs. The reason for this discrepancy was the coastal location in the case of the Acaraù station (Figure 1) and the wind speed value in the case of the Campos Sales and Crateùs stations (Table 2), confirming the results obtained by Fooladmand and Haghighat (2007).

4.2 SPATIALIZATION

The spatialization of the adjusted coefficients in both calibrations (Figure 4) allowed for a better understanding of the behaviour and sensitivity of the parameterized values. The simultaneous adjustment of the α and β parameters shows that they behave inversely proportionally, with the alpha parameter oscillating between 0.0005 and 0.0023.

Near the Guaramiranga station, the alpha coefficient shows its highest value. This result can be attributed to the low air temperatures recorded in the region, due to the high altitude (Table 1 and 2). The same behavior occurs in the far east of the state, in the vicinity of the Apodi plateau, and the possible cause of this representation is not altitude, but the lack of a station located in the highest part of the region. The conventional weather station in Apodi - RN is located at an altitude of only 150 m, which does not represent the geographic formation of the plateau, which certainly has an influence on the region's climate. In the rest of the state, the α coefficient decreases as you move away from the mountainous region and/or as a result of the higher wind speeds and air temperatures recorded in the rest of the state.

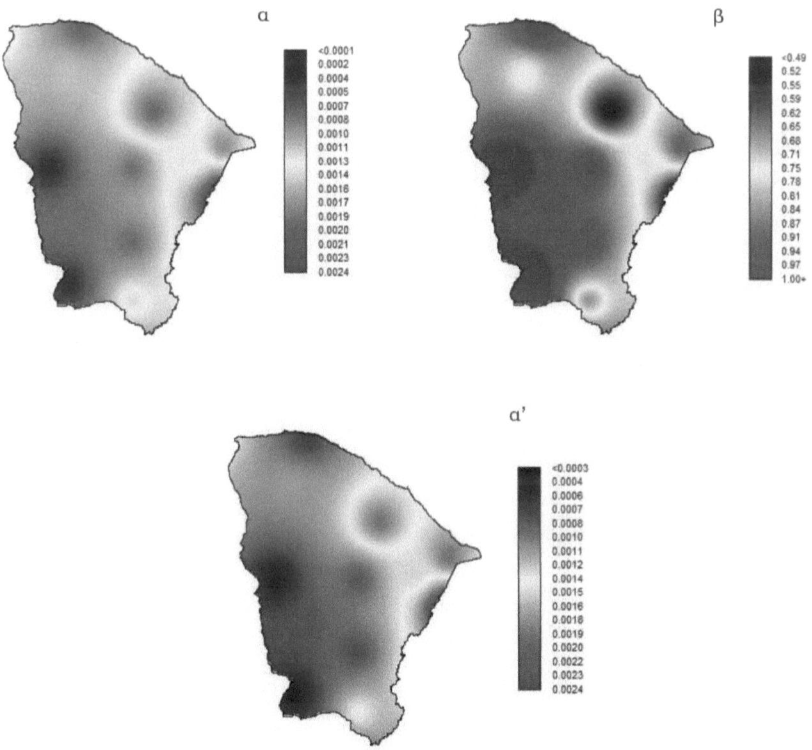

Figure 4 - Spatial distribution of the alpha and beta coefficients, calibrated simultaneously and the adjustment made only with the alpha' coefficient

The β parameter oscillated between 0.49 and 1.0 approximately, with its highest values in almost every state. As it is the exponent of the temperature difference in the HS model, thus representing the temperature range, the beta exponent shows greater sensitivity, thus expressing its greater importance among the adjusted parameters, given that the study region has a well-defined climate, classified as semi-arid, with high temperatures almost all year round. The spatialization made with the model in which only the alpha coefficient α' was adjusted ranged from 0.0004 to 0.0022 approximately, behaving similarly to the spatialization of alpha in the simultaneous adjustment.

The spatialization of reference evapotranspiration allowed us to understand the spatial and temporal variability of the entire state of Ceará (Figures 5 to 16). Xu *et al.* (2006) explain that spatial distribution maps provide very valuable information for the management and planning of water resources in a basin or region, due to the great importance of ETo as an element of the hydrological cycle.

In the peak ETo months, the location that showed the lowest evapotranspiration demand for modeling with the original HS equation and with HScal(α/β) was the Guaramiranga mountain region, which once again can be explained by its high altitude and low air temperature normals, while in the months of March and May (Figure 7 and 9) the lowest ETo demand can be observed on the state coast. In the model where only the α' coefficient was adjusted, the sites with the lowest evapotranspiration demands are located in the west of the state, where the modeling is influenced by the adjustments made in the cities of Crateùs and Campos Sales and the interpolations with stations in the state of Piaui, showing values that are very different from those observed in the other two models.

By analyzing the spatial and temporal variation of ETo estimates (Figures 5 to 16), it is possible to understand the behavior of the three models throughout the state of Ceará. The original model and the calibration made with α and β showed their highest evapotranspiration demands in the month of October, approximately 8.0 mm for the original HS and 10.0 mm for Hscal (α/β). in the far west of the region studied.

For the month with the highest evapotranspiration demand, high ETo values were expected, but in the region of 8.0 or 9.0 mm. This behaviour in estimating the spatialization of the calibration of α and β can be explained by the lack of representativeness of the stations located in the extreme east and west of the state evaluated, since on the borders between Cearà and Piaui there is a large geographical elevation of over 700 m (Chapada da Ibiapaba), And there is no data collection station at this high altitude, so spatialization in that region may have been hampered by the lack of representativeness of the interpolation of estimates from the northwest of the state of Cearà with those from the east of the state of Piaui. Similar results were found by Oliveira *et al.* (2013).

The spatialization made with the model that was adjusted only for the α coefficient underestimated the other models studied in all the months of the year, and because the β coefficient was not modified, which in turn represents the sensitivity of the temperature range in the HS equation, its spatialization showed a high disparity throughout the period studied, estimating a maximum of 7.00 mm, and in a completely opposite region to where the other models estimated the ETo peaks. Tabari *et al.* (2013) spatialized reference evapotranspiration in southeastern Ira (the country's most arid region) using the adjusted Blaney-Criddle equation and found the highest ETo estimates (11.0 mm) for the months of June, July and August.

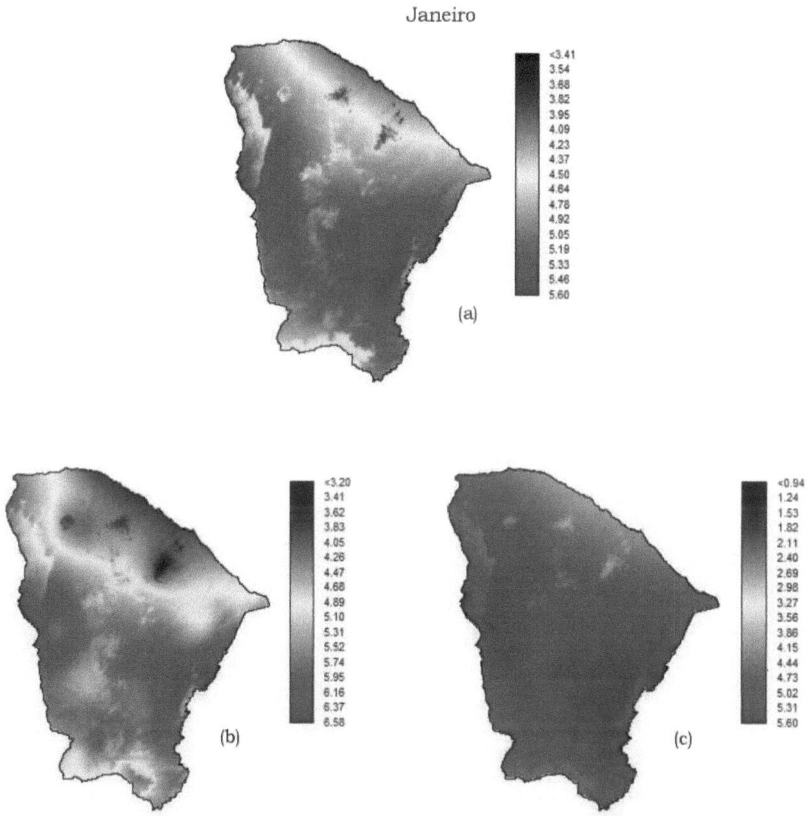

Figure 5 - Spatial distribution of ETo (mm day^{-1}) by Original Hargreaves (a); Hargreaves with simultaneous adjustment of the α and β parameters (b) and Hargreaves with adjustment of the alpha coefficient only (c) for the month of January.

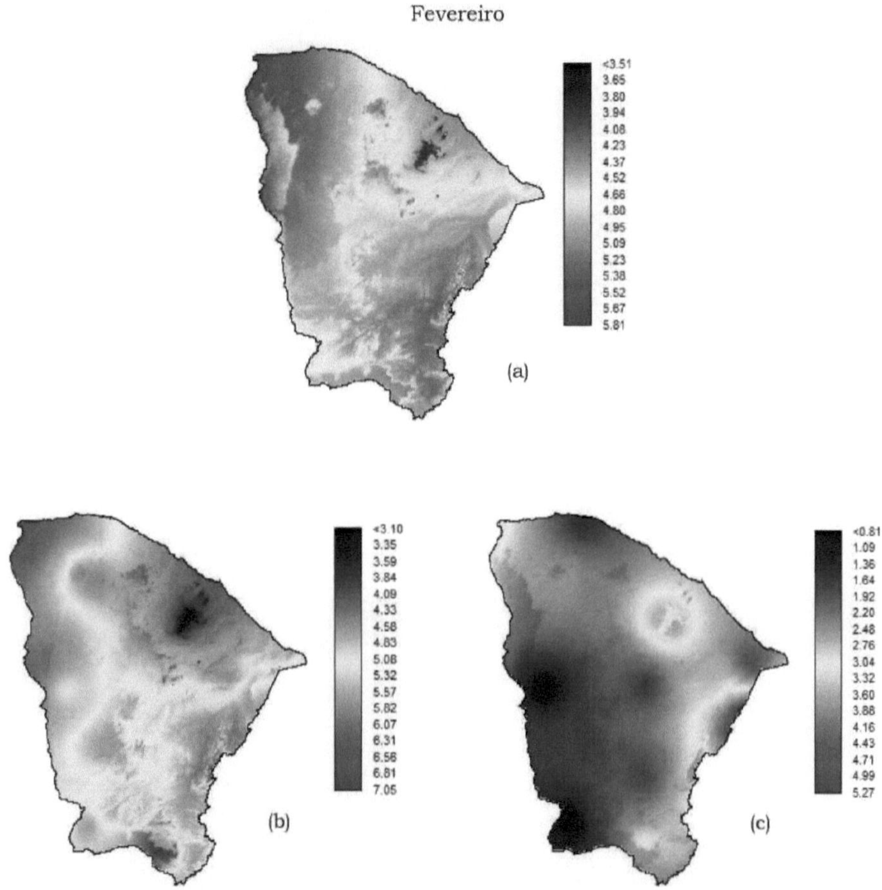

Figura 6 - Spatial distribution of ETo (mm day^{-1}) by Original Hargreaves (a); Hargreaves with simultaneous adjustment of the α and β parameters (b) and Hargreaves with adjustment of the α' coefficient only (c) for the month of February.

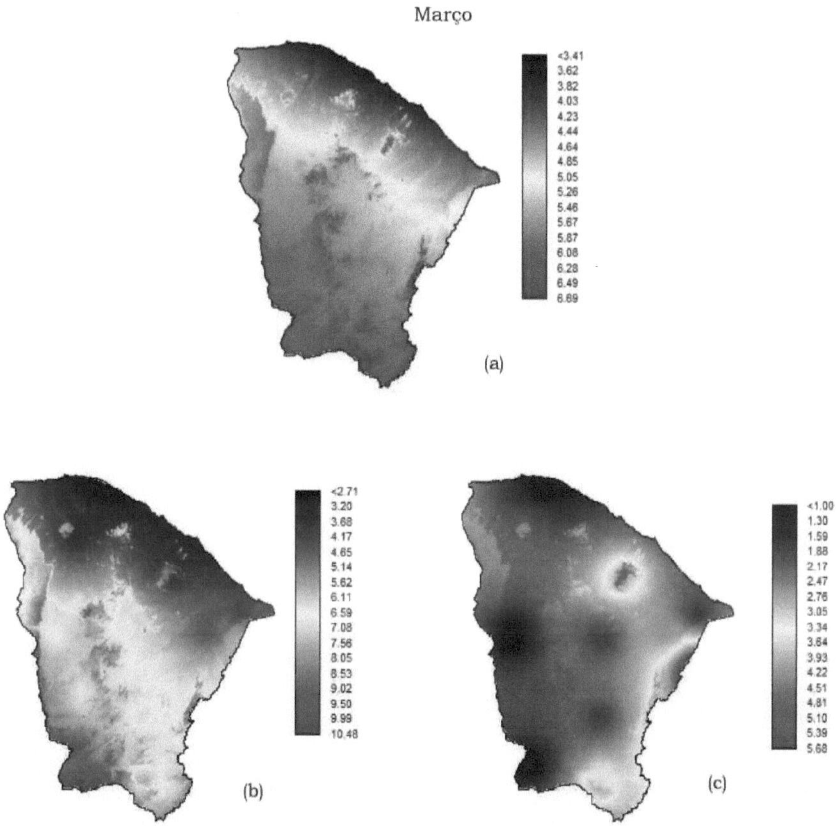

Figura 7 - Spatial distribution of ETo (mm day^{-1}) by Original Hargreaves (a); Hargreaves with simultaneous adjustment of the α and β parameters (b) and Hargreaves with adjustment of the α' coefficient only (c) for the month of March.

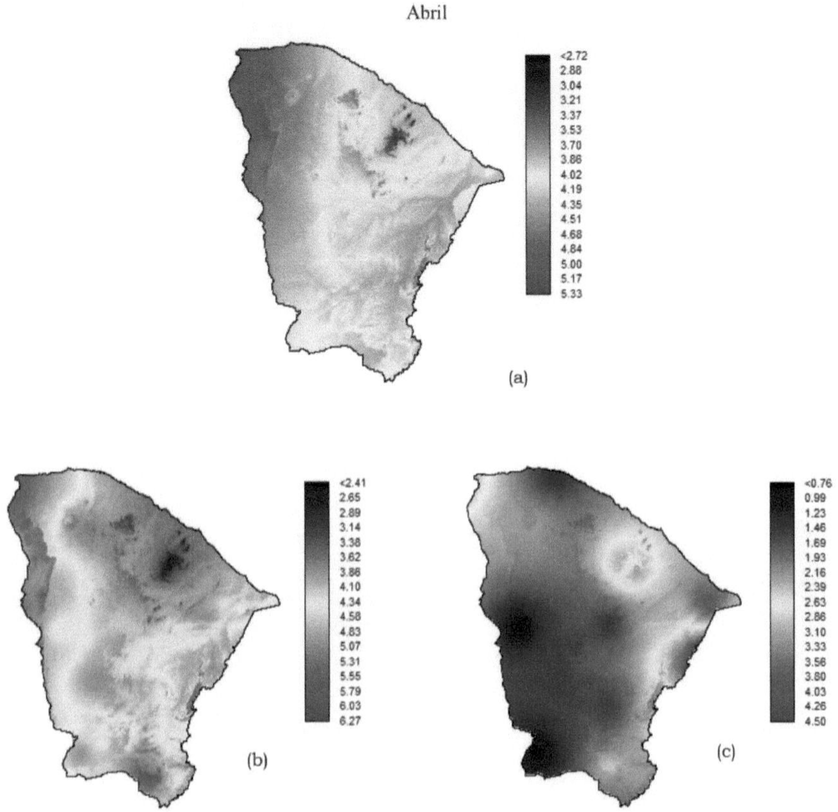

Figura 8 - Spatial distribution of ETo (mm day-1) by Original Hargreaves (a); Hargreaves with simultaneous adjustment of the α and β parameters (b) and Hargreaves with adjustment of the α' coefficient only (c) for the month of April.

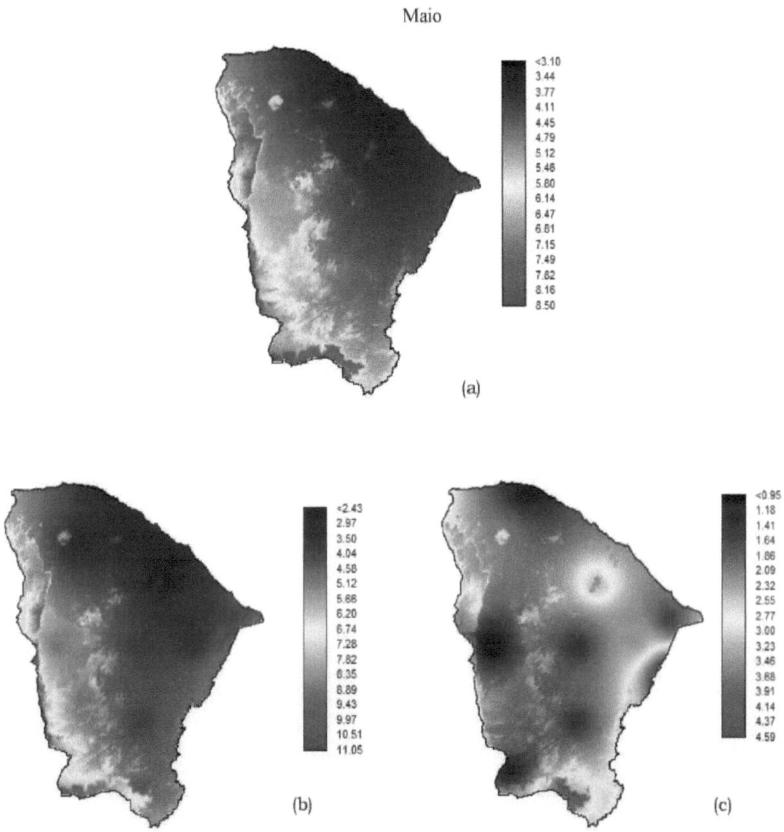

Figure 9 - Spatial distribution of ETo (mm day^{-1}) by Original Hargreaves (a); Hargreaves with simultaneous adjustment of the α and β parameters (b) and Hargreaves with adjustment of the α' coefficient only (c) for the month of May.

Junho

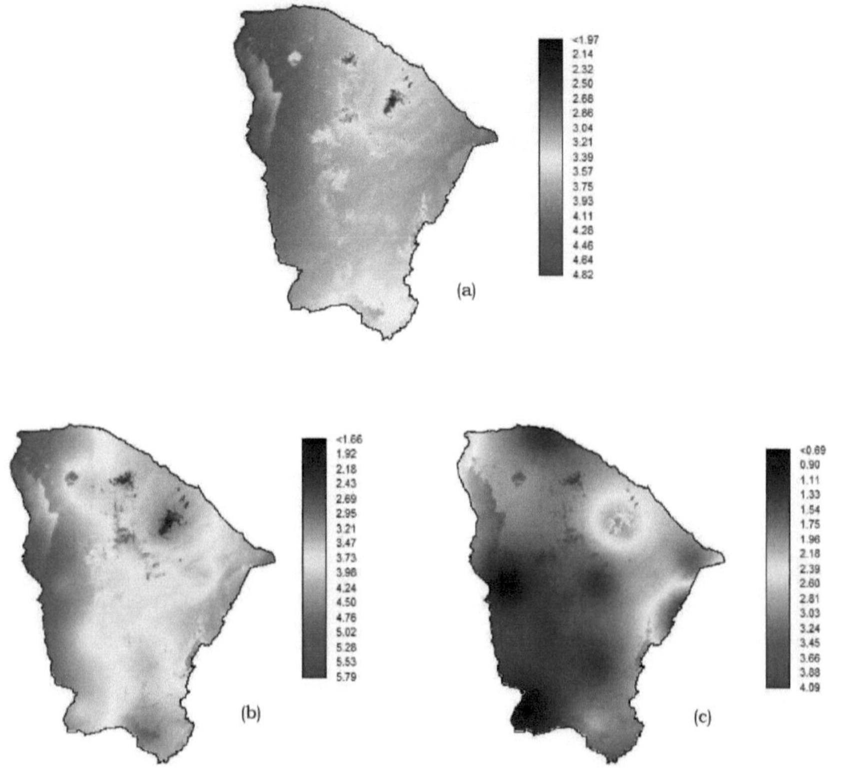

Figure 10 - Spatial distribution of ETo (mm day^{-1}) by Original Hargreaves (a); Hargreaves with simultaneous adjustment of the α and β parameters (b) and Hargreaves with adjustment of the α' coefficient only (c) for the month of June.

Julho

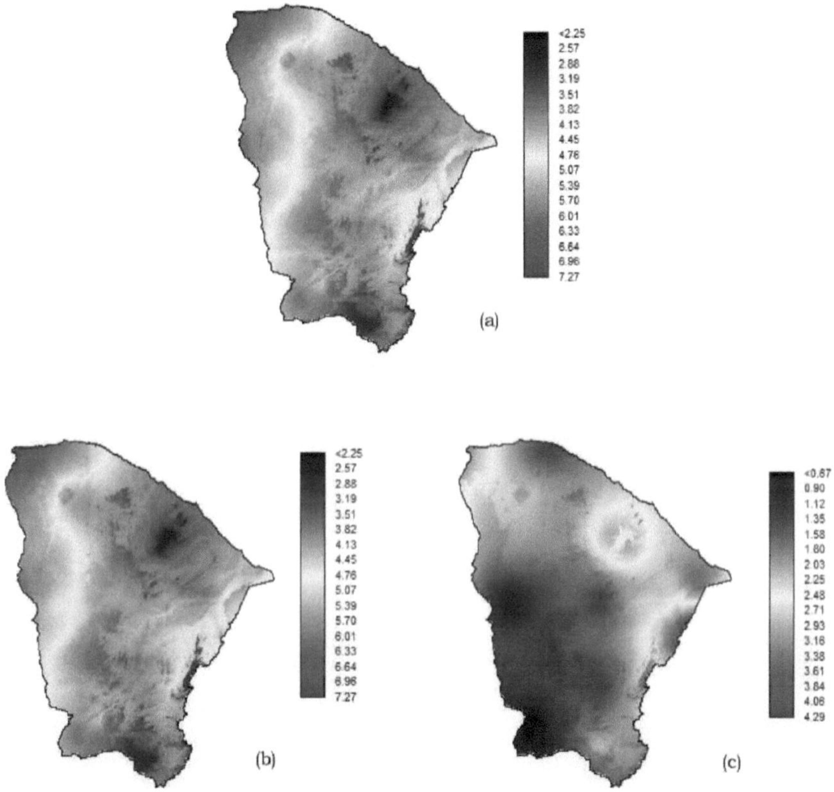

Figure 11 - Spatial distribution of ETo (mm day^{-1}) by Original Hargreaves (a); Hargreaves with simultaneous adjustment of the α and β parameters (b) and Hargreaves with adjustment of the α' coefficient only (c) for the month of July.

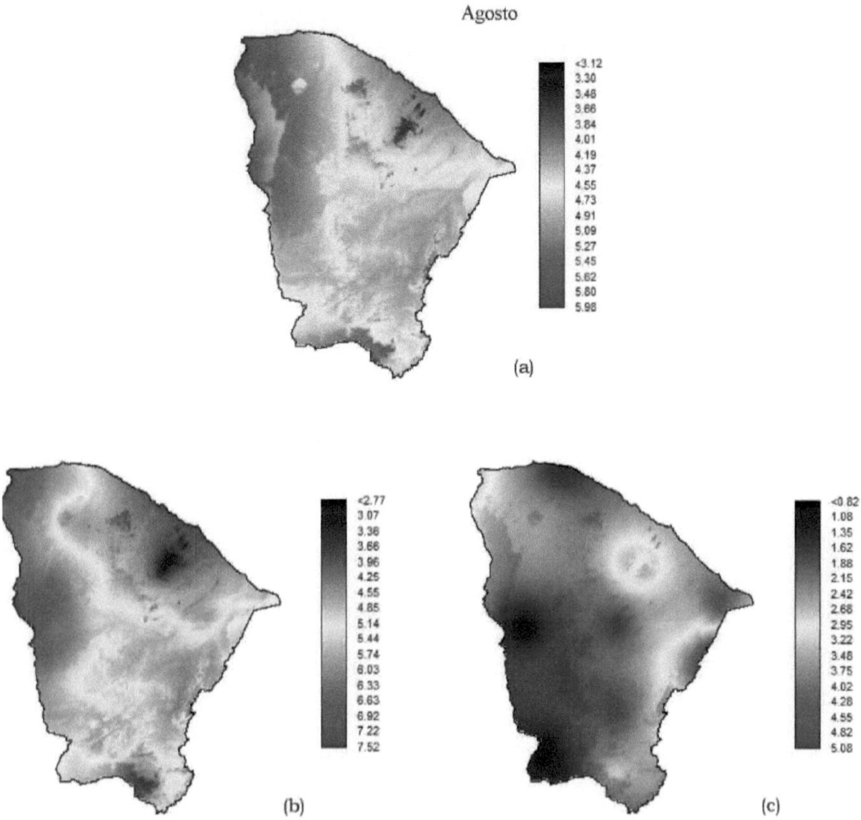

Figure 12 - Spatial distribution of ETo (mm day^{-1}) by Original Hargreaves (a); Hargreaves with simultaneous adjustment of the α and β parameters (b) and Hargreaves with adjustment of the α' coefficient only (c) for the month of August.

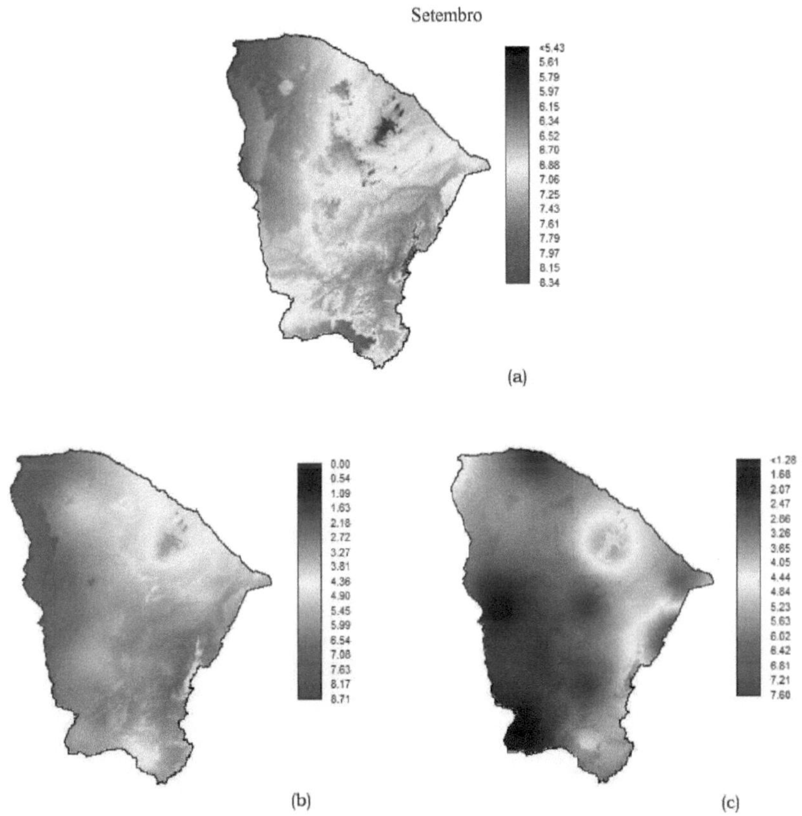

Figure 13 - Spatial distribution of ETo (mm day^{-1}) by Original Hargreaves (a); Hargreaves with simultaneous adjustment of the α and β parameters (b) and Hargreaves with adjustment of the α' coefficient only (c) for the month of September.

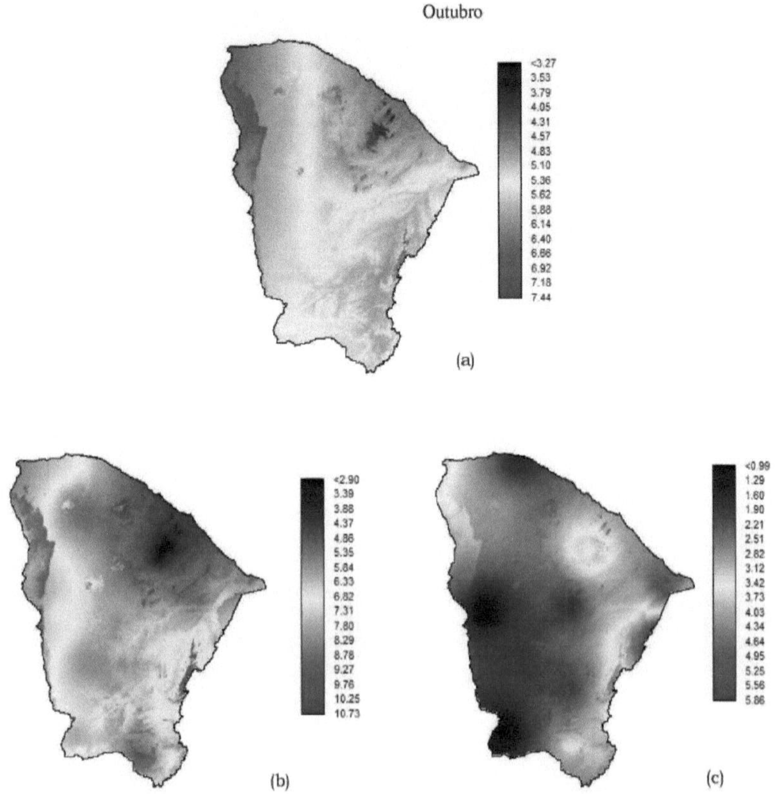

Figure 14 - Spatial distribution of ETo (mm day^{-1}) by Original Hargreaves (a); Hargreaves with simultaneous adjustment of the α and β parameters (b) and Hargreaves with adjustment of the α' coefficient only (c) for the month of October.

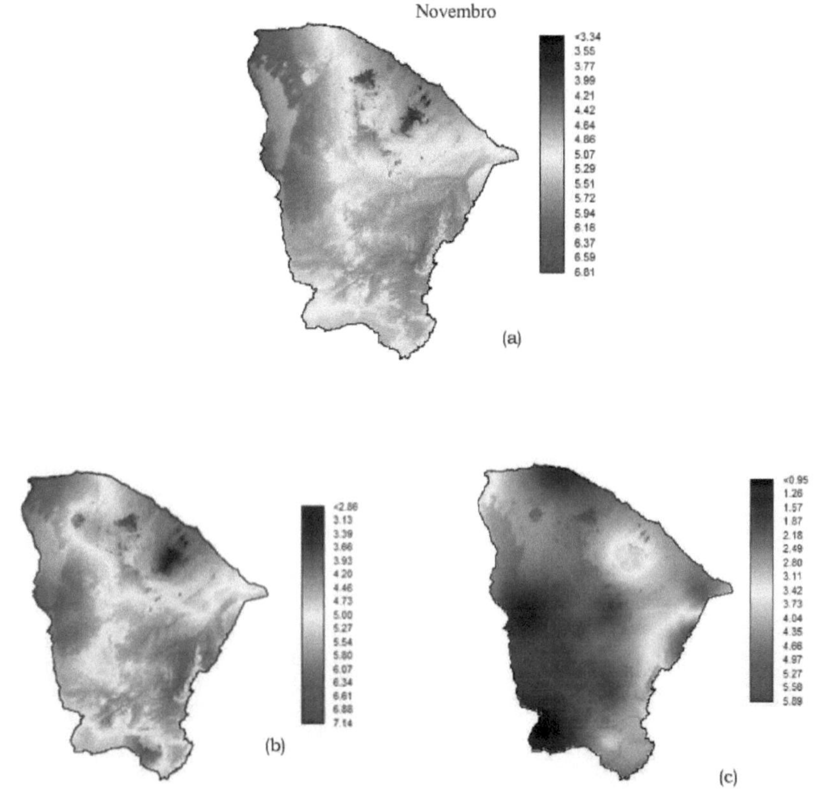

Figure 15 - Spatial distribution of ETo (mm day^{-1}) by Original Hargreaves (a); Hargreaves with simultaneous adjustment of the α and β parameters (b) and Hargreaves with adjustment of the α' coefficient only (c) for the month of November.

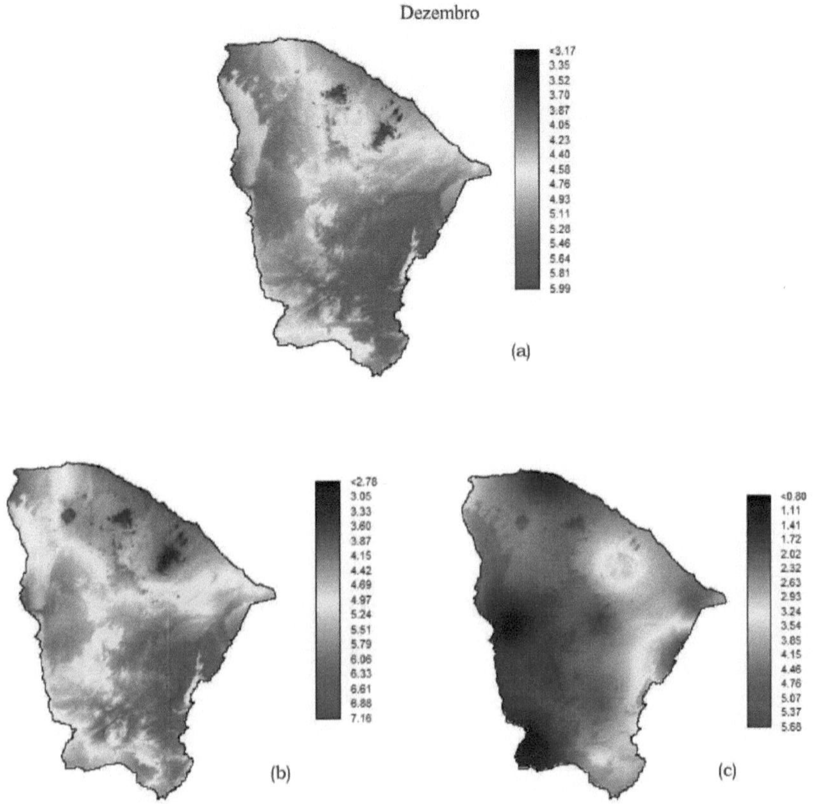

Figure 16 - Spatial distribution of ETo (mm day^{-1}) by Original Hargreaves (a); Hargreaves with simultaneous adjustment of the α and β parameters (b) and Hargreaves with adjustment of the α' coefficient only (c) for the month of December.

Figures 17 to 20 show the comparisons between the estimates made with the original and calibrated coefficients of the Hargraves and Samani (1985) equation, clearly showing the spatial and temporal variation in ETo estimates, the lowest evapotranspiration demands occurred in Guaramiranga, which reinforces the results obtained in the calibration and spatialization of ETo for Ceará. In both analyses, Guaramiranga is strongly presented as the location with the lowest ETo demand in the PM, HS and HScal(α/β) equations. It is important to note that the HS equation overestimates ETo in regions with very high humidity, and underestimates it in very dry regions and with higher wind speeds, because in its original modeling they were unable to include the wind speed term (u_2) satisfactorily (XU; SINGH, 2002; HARGRAVES; ALLEN, 2003).

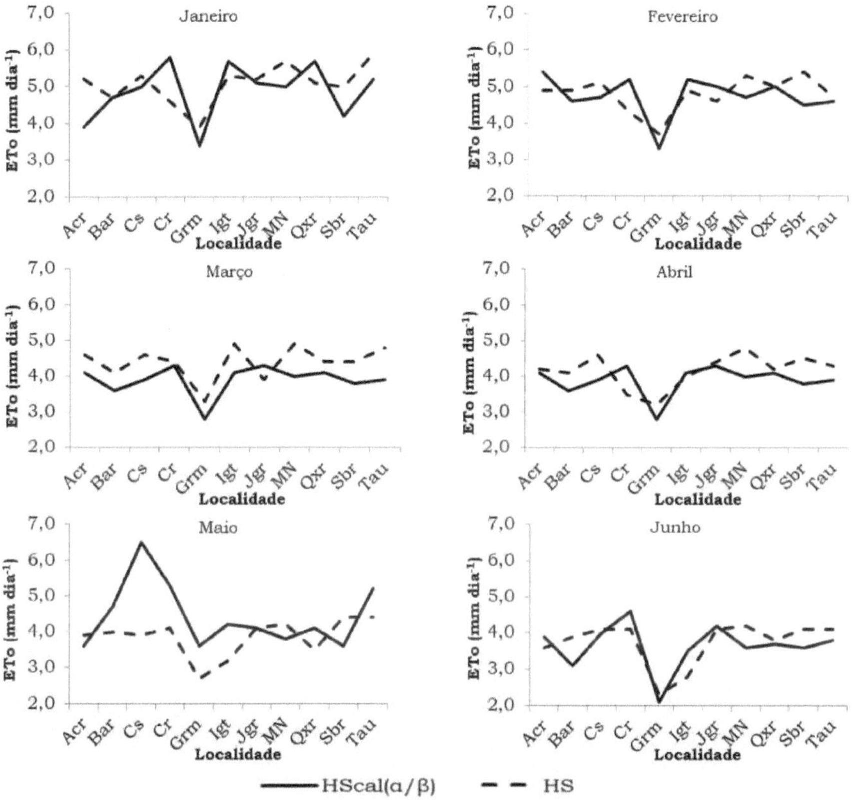

Figura 17 - Analysis of the ETo trend obtained from the map using the original Hargreaves and Samani equation and with the alpha and beta coefficients calibrated for the first half of the year.

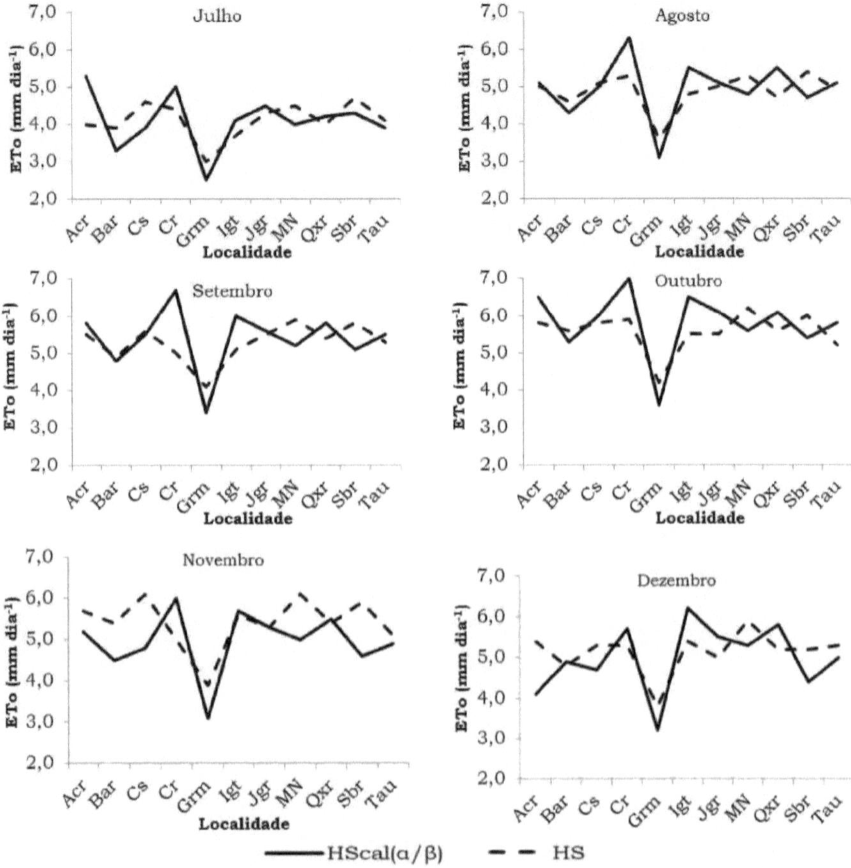

Figura 18 - Analysis of the ETo trend obtained from the map using the original Hargreaves and Samani equation and with the alpha and beta coefficients calibrated for the second half of the year.

Analyzing the variation in the ratio between the averages (Table 6), which oscillated between 89.5 and 110.6% for the months of May and November respectively, it can be seen that in the month of May the original equation underestimated (10.5%) the results obtained with the calibrated model, Given that only in the months of August and September did the equations compared hardly differ from each other, months in which the highest temperatures routinely occur in the state of Cearà, and as the HS model is based on temperature, it suffered little change in the adjustment made. The same table also shows that the EPE varied from 0.54 in February and April to 1.11 in May. The EPM ranged from 9.17 (February) to 22.1 (May), results which emphasize the greater difference between the equations used in May.

Table 5 - Mean ratio (ξ), root mean square error (RMSE) and mean percentage error (EPM) between ETo estimated by digital elevation models using the original HS equations (1985) and HS calibrated with alpha and beta.

Month	ξ (%)	RMSE (mm day)$^{-1}$	EPM(%)
Jan	105,8	0,75	11,90
Feb	102,0	0,54	9,17
Sea	113,0	0,64	12,72
Apr	107,4	0,54	10,35
May	89,5	1,11	22,61
Jun	103,5	0,47	10,55
Jul	102,5	0,62	12,58
Aug	99,9	0,57	9,39
Set	99,4	0,74	10,45
Out	97,3	0,70	11,19
Nov	110,6	0,86	12,26
Ten	105,2	0,70	11,67

The comparisons between the estimates made with the HS equation, which was adjusted only for the alpha coefficient, and HS with the original coefficients (Figures 19 and 20), again clearly show the spatial and temporal variation in ETo estimates, a spatial variation that is contrary to the behavior of ETo in the state of Ceará, as pointed out by the calibration carried out in this work, as well as the behavior shown by Oliveira *et al* (2013), with the lowest evapotranspiration demands occurring in a totally different region to those found by the simultaneous calibration, being recorded in Campo Sales and Crateùs.

In the calibration with only α', the original equation underestimated the reference evapotranspiration in practically every month, except for May in Guaramiranga, when it recorded its highest estimate, which was approximately 3.5 mm. This reinforces the results obtained in the calibration and spatialization of ETo for Ceará. In both analyses, Guaramiranga emerges strongly as the location with the lowest ETo demand in the PM, HS and calibrated HS α and β equations.

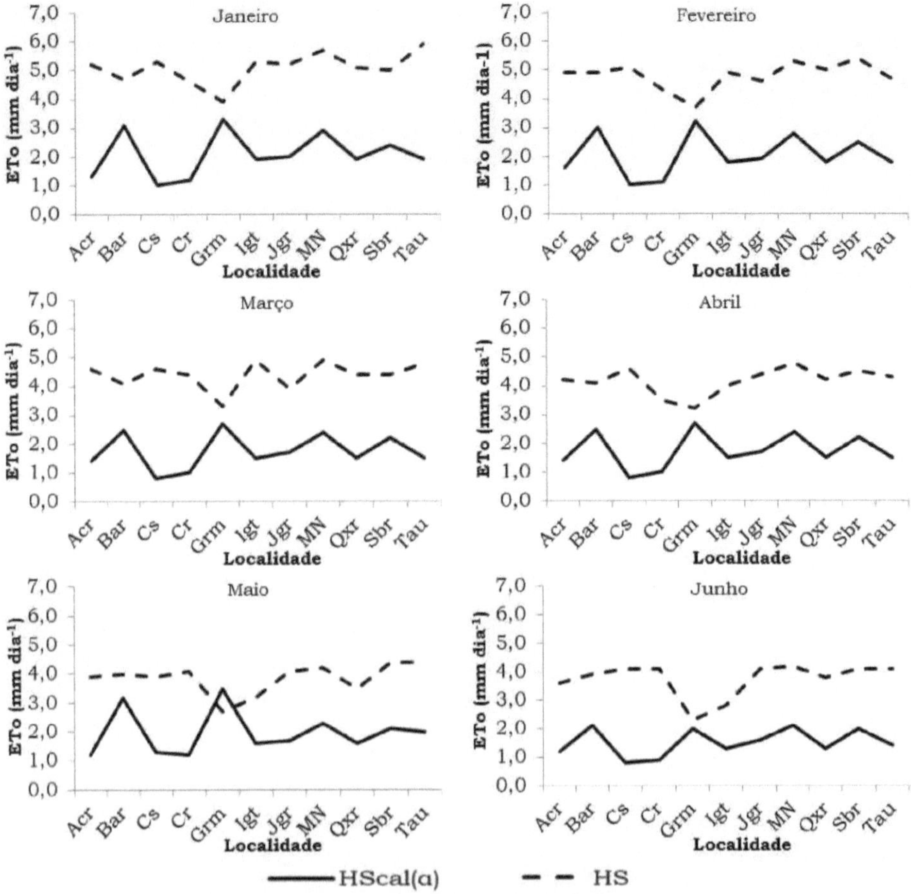

Figura 19 - Analysis of the ETo trend obtained from the map using the original Hargreaves and Samani equation and with the alpha coefficient calibrated for the first half of the year.

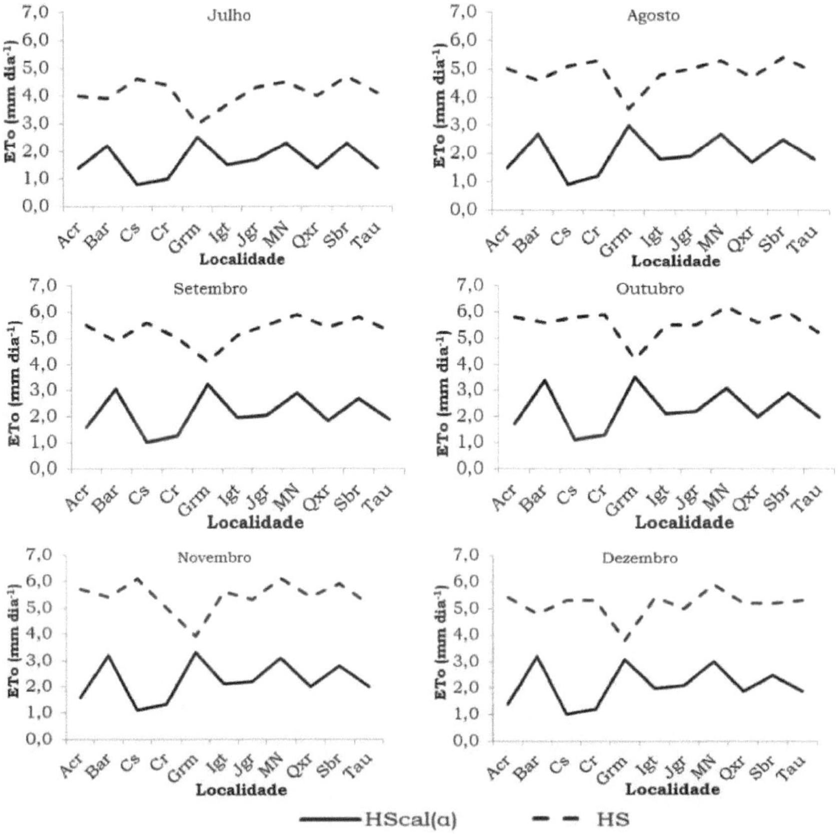

Figura 20 - Analysis of the ETo trend obtained from the map using the original Hargreaves and Samani equation and with the alpha coefficient calibrated for the second half of the year.

The analysis of the ratio between the mean estimates between HS (α') and the original HS (Table 7) shows the high disparity, with the ratio between the means ranging from 221.9 (May) to 291.2% (March). The same table also shows that the EPE reached very high values, ranging from 2.24 to 3.61 for the months of May to October respectively. The EPM oscillated between an exorbitant 51.92 (May) and 58.92 (March), results which show very clearly that the calibration which adjusted only the alpha coefficient (α) of the Hargraves and Samani equation, which in its original form is 0.0023, does not satisfactorily estimate the reference evapotranspiration for the state of Ceará.

Table 6 - Ratio of means (ξ), standard error of estimate (RMSE) and average percentage error (EPM) between the ETo estimated by the digital elevation model using the original HS equations (1985) and HS with calibration of the alpha parameter only.

Month	ξ (%)	RMSE(mm day)$^{-1}$	EPM(%)
Jan	282,3	3,33	57,89
Feb	267,7	3,03	56,64
Sea	291,2	2,93	58,92
Apr	273,1	2,67	57,25
May	221,9	2,24	51,92
Jun	275,5	2,47	57,44
Jul	279,7	2,69	57,81
Aug	283,9	3,21	58,37
Set	282,7	3,45	58,66
Out	278,7	3,61	57,72
Nov	274,8	3,50	57,32
Ten	281,4	3,35	57,63

5 CONCLUSION

The calibration of the coefficients of the Hargreaves and Samani equation for the state of Ceará showed satisfactory results in estimating reference evapotranspiration in places where only maximum and minimum temperature data is available. The adjustment that showed the best results was the simultaneous calibration of the α coefficient and the β exponent.

The spatialization of evapotranspiration demand in the state of Ceará proved feasible when the simultaneously calibrated parameters of the Hargraves and Samani (1985) equation were used. The equation in which only the α coefficient was adjusted did not show satisfactory results at any time of the year in most of the state.

REFERENCES

ALLEN, R. G. *et al.* An update for the definition of reference evapotranspiration. **ICID Bull**, v. 43, n. 2, p.1-34, 1994

ALLEN, R. G. *et al.* **Crop evapotranspiration**: guidelines for computing water requirements. Rome: FAO, 1998. 300 p. (FAO, Irrigation and Drainage Paper, 56)

ANDRÉ, R. G. B. et al. Energy aspects of sugarcane development. Part 1: radiation balance and derived parameters. Revista Brasileira de Meteorologia, v. 25, n. 3, p. 375-382, 2010.

BAUTISTA, F.; BAUTISTA, D.; DELGADO-CARRANZA, C. Calibration of the equations of Hargreaves and Thornthwaite to estimate the potential evapotranspiration in semi-arid and subhumid tropical climates for regional applications. Atmosphere, v. 22, p. 331-348, 2009.

BARDIN, L.; PEDRO JÛNIOR, M. J. ; MORAIS, J. F. L. Estimation of maximum and minimum air temperatures for the region of Circuito das Frutas, SP. **Revista Brasileira de Engenharia Agricola e Ambiental**, v. 14, n. 6, p. 618-624, 2010.

BERNARDO, S.; SOAREA, A. A.; MANTOVANI, E. C. Manual de irrigaçao: Irrigation Manual. 8 ed. Viçosa-MG: UFV, 2006. 611

BORGES, A. C.; MENDIONDO, E. M. Comparison between empirical equations for estimating reference evapotranspiration in the Jacupiranga River Basin. **Revista Brasileira de Engenharia Agricola e Ambiental**, v.11, n.3, p.293-300, 2007.

CARGNELUTTI FILHO, A.; MALUF, J. R. T.; MATZENAUER, R. Geographical coordinates in the estimation of maximum and average decadal air temperatures in the State of Rio Grande do Sul. **Ciência Rural**, v. 38, n. 9, p. 2448-2456, 2008.

DEHGHANISANIJ, H.; YAMAMOTO, T.; RASIAH, V. Assessment of evapotranspiration estimation models for use in semi-arid environments. **Agricultural Water Management**, v. 64, n. 2, p. 91-106, 2004.

DROOGERS P, ALLEN R. Estimating reference evapotranspiration under inaccurate data conditions. **Irrigation and Drainage Systems** 16 (1): 33-45, 2002.

FERNANDES, D.S. *et al.* Regional and local calibration of the Hargreaves equation for estimating reference evapotranspiration. **Revista Ciência Agronômica**, v. 43, n. 2, p. 246-255, Apr-Jun, 2012

FOOLADMAND, H. R.; ZANDILAK, H.; RAVANAN, M. H. Comparison of different types of Hargreaves equation for estimating monthly evapotranspiration in the south of Iran. Archives of Agronomy and Soil Science, v. 54, n. 3, p. 321-330, 2008.

FOOLADMAND, H.R.; HAGHIGHAT, M. Spatial and temporal calibration of Hargreaves equation for calculating monthly ETo based on Penman-Monteith method. **Irrigation and Drainage**, v.56, p 439-444, 2007.

FONTANE, D. "Multi-objective simulation andoptimization of reservoir operation using Excel." Proceeding of the Second Federal Interagency Hydrologic Modeling Conference, Las Vegas, NV, USA, 2001.

GAVILAN, P. *et al.* Regional calibration of Hargreaves equation for estimating reference ET in a semi-arid environment. **Agricultural Water Management**, v. 81, n. 3, p. 257-281, 2006.

GAVILAN, P.; CASTILLO-LLANQUE, F. Estimating reference evapotranspiration with atmometers in a semiarid environment. Agricultural Water Management, v. 96, n. 3, p. 465-472, 2009.

HARGREAVES, G. H.; ALLEN, R. G. History and evaluation of Hargreaves evapotranspiration equation. Journal of Irrigation and Drainage Engineering, v. 129, n. 1, p. 53-63, 2003.

HARGREAVES, G. H.; SAMANI, Z. A. Estimating potential evapotranspiration. **Journal of Irrigation and Drainage Engeneering**, v. 108, p. 225-230, 1982.

HARGREAVES, G. H.; SAMANI, Z. A. Reference crop evapotranspiration from temperature. **Applied Engineering in Agriculture**, v. 01, n. 02, p. 96-99, 1985.

HENGGELER J, SAMANI Z, FLYNN M, ZEITLER J. 1996. Evaluation of various evapotranspiration equations for Texas and New Mexico. Proceeding of Irrigation Association International Conference,San Antonio, Texas.

HUDDLESTON, D. H., ALARCON, V. J., AND CHEN, W. "Water distribution network analysis using Excel." **Journal of Hydraulic Engineering**, ASCE, Vol. 130, No. 10, pp. 1033-1035, 2004.

IRMAK, S. *et al.* Solar and Net radiation-based equation to estimate reference evapotranspiration in humid climates, **Journal of Irrigation and Drainage Engineering**, v. 129, n. 5, p. 360-370, 2003.

ITENFISU, D. *et al.* Comparison of reference evapotranspiration calculations as part of the ASCE standardization effort, **Journal of Irrigation and Drainage Engineering,** v. 129, n. 6, p. 440-448, 2003.

JERSZURKI, D.; SOUZA, J. L. M. Parameterization of the Hargreaves & Samani and Angstron-Prèscott equations for estimating solar radiation in the region of Telêmaco Borba, State of Paranà. **Ciência Rural**, v. 43, n. 3, p. 383-389, 2013.

LEGATES, D.R.; MCCABE, G.J. Evaluating the use of "goodness-of-fit" measures in hydrologic and hydro climatic model validation. **Water Resources Research**. v. 35, n. 1, p. 233-241, 1999.

LUO, J. AND XIE, J. "Parameter estimation for nonlinear

Muskingum model based on immune clonal selection algorithm." **Journal of Hydrologic Engineering**, ASCE, Vol. 15, No. 10, pp. 844-851, 2010.

MAESTRO-VALERO, J. F. *et al.* Regionalization of the Hargreaves coefficient to estimate long-term reference evapotranspiration series in SE Spain. **Spanish Journal of Agricultural Research**, v.11, n.4, p.1137-1152, 2013.

MARTINEZ-COB, A.; TEJERO-JUSTE, M. A wind-based qualitative calibration of the Hargreaves ET estimation equation in semiarid regions. Agricultural Water Management, v. 64, n. 3, p. 251-264, 2004. ISSN

0378-3774.

MELO, G. L. de; FERNANDES, A. L. T. Evaluation of empirical methods to estimate reference evapotranspiration in Uberaba, State of Minas Gerais, Brazil. **Engenharia Agricola**, Jaboticabal, v.32, n.5, p.875-888. 2012.

OLIVEIRA, J. B; ARRAES, F. D. D. A methodology for the spatialization of reference evapotranspiration from SRTM data. **Revista Ciência Agronômica**, v. 44, n. 3, p. 445454, jul/set. 2013.

RAZIEI, T.; PEREIRA, L. S. Estimation of ETo with Hargreaves-Samani and FAO- PM temperature methods for a wide range of climates in Iran. **Agricultural Water Management**, v. 121, p. 1-18, 2013.

REICHARDT, K.; TIMM, L. C. Soil, plant and atmosphere: concepts, processes and applications: Soil, plant and atmosphere: concepts, processes and applications: **Manole**, 2004

ROCHA, É. J. T. *et al.* Estimation of ETo by the Penman-Monteith FAO model with minimum data integrated with a Geographic Information System. **Revista Ciência Agronômica**, v. 42, n. 1, p. 75-83, 2011.

SABZIPARVAR, A. A.; TABARI, H. Regional estimation of reference evapotranspiration in arid and semiarid regions. **Journal of Irrigation and Drainage Engineering**, v. 136, n. 10, p. 724-731, 2010.

SENTELHAS, P. C.; GILLESPIE, T. J.; SANTOS, E. A. Evaluation of FAO Penman- Monteith and alternative methods for estimating reference evapotranspiration with missing data in Southern Ontario, Canada. **Agricultural Water Management**, v. 97, n. 5, p. 635-644, 2010.

SEPASKHAH, A. R.; RAZZAGHI, F. Evaluation of the adjusted Thornthwaite and Hargreaves-Samani methods for estimation of daily evapotranspiration in a semi-arid region of Iran. **Archives of Agronomy and Soil Science**, v. 55, n. 1, p. 51-66, 2009.

SHAHIDIAN, S. *et al.* Parametric calibration of the Hargreaves-Samani equation for use at new locations hydrological processes. **Hydrological Process**. 27, 605-616 2013

TABARI, H.; TALAEE, P. H. Local Calibration of the Hargreaves and Priestley-Taylor Equations for Estimating Reference Evapotranspiration in Arid and Cold Climates of Iran Based on the Penman-Monteith Model. **Journal of Hydrologic Engineering**, v. 16, n. 10, p. 837-845, 2011.

TABARI, H.; TALAEE, P. H.; SOME B. S. Spatial modelling of reference evapotranspiration using adjusted Blaney-Criddle equation in an arid environment, **Hydrological Sciences Journal**, DOI:10.1080/02626667.2012.755265. 2013.

TALAEE, P. H. Performance evaluation of modified versions of Hargreaves equation across a wide range of Iranian climates. **Meteorology and Atmospheric Physics**. p. 16, 2014.

TEMESGEN, B. *et al.* Adjusting temperature parameters to reflect well-water conditions. **Journal of Irrigation and Drainage Engineering**, v. 125, n. 1, p. 26-33, 1999.

THEPADIA, M.; MARTINEZ, C. J. Regional Calibration of Solar Radiation and Reference Evapotranspiration Estimates with Minimal Data in Florida. **Journal of Irrigation and Drainage**

Engineering, v. 138, n. 2, p. 111-119, 2012.

TRAJKOVIC, S. Hargreaves versus Penman-Monteith under humid conditions.

Journal of Irrigation and Drainage Engineering, v. 133, n. 1, p. 38-42, 2007.

WRAITH, JON M.; OR, D. Nonlinear parameter estimation using spreadsheet software.

Journal Natural Resources and Life Science Education. v. 27: p. 13-19, 1998.

XU C.Y.; SINGH, V. P. Cross comparison of empirical equations for calculating potential evapotranspiration with data from Switzerland. **Water Research Management**, v. 16, p. 197-219, 2002.

XU, C.Y *et al.* Analysis of spatial distribution and temporal trend of reference evapotranspiration and pan evaporation in Changjiang (Yangtze River) catchment.

Journal of Hydrology, v. 327, n. 1/2, p. 81- 93, 2006.

MIX
Papier aus verantwortungsvollen Quellen
Paper from responsible sources
FSC® C105338

Printed by Books on Demand GmbH, Norderstedt / Germany